TIMBER POLE CONSTRUCTION

LIONEL JAYANETTI

Practical
ACTION
PUBLISHING

Intermediate Technology Publications 1990

Practical Action Publishing Ltd
25 Albert Street, Rugby, CV21 2SD, Warwickshire, UK
www.practicalactionpublishing.com

© Intermediate Technology Publications 1990

First published 1990\Digitised 2013

ISBN 10: 1 85339 502 1
ISBN 13 Paperback: 9781853395024
ISBN Library Ebook: 9781780441566
Book DOI: http://dx.doi.org/10.3362/9781780441566

Since 1974, Practical Action Publishing has published and disseminated
books and information in support of international development work
throughout the world. Practical Action Publishing is a trading name
of Practical Action Publishing Ltd (Company Reg. No. 1159018), the
wholly owned publishing company of Practical Action. Practical Action
Publishing trades only in support of its parent charity objectives and any
profits are covenanted back to Practical Action (Charity Reg. No. 247257,
Group VAT Registration No. 880 9924 76).

CONTENTS

1. BACKGROUND

Introduction

Timber poles have been one of the most valuable building materials throughout history. In the modern world, forest plantations are raised with trees planted in close proximity to each other so that, in their early stages, they grow up in slim upright manner, with little development of lateral branches. Since the full potential of the site is concentrated on a fewer number of trees per hectare forming the final crop, thinnings are done at intervals of time. These thinnings often provide trees of a quite reasonable size, which could be efficiently used in the round form. In most cases, these timber poles have a fairly thick outer sapwood layer which is easily penetrated by preservatives, and it is sufficient to provide a continuous and reasonably thick outer layer of protected wood. As a result of the establishment of community forestry programmes in many parts of the world, more poles will be available in the future than at present.

Advantages of pole construction

A timber pole is stronger than sawn timber of equal cross-sectional area, because fibres flow smoothly around natural defects and are not terminated as sloping grain at cut surfaces.

A round pole possesses a very high proportion of the basic strength of its species, because knots have less effect on the strength of naturally round timbers, compared with sawn sections.

The cost and wastage of sawing are eliminated.

Sawn timber is produced from trees of large diameters which have grown for several decades; their replacement takes a longer time than that for poles which take less time to grow; and felling the former in large numbers can cause serious environmental problems.

Above all, the design of any pole structure can be simple enough for unskilled persons to construct.

Species

Most common species of trees will provide poles which can be utilized for many purposes and those of perishable species can be preserved to give them a life as long or longer than untreated durable species. It has been found in many developing countries that poles from mangrove swamps, thinnings from eucalyptus or softwood plantations, etc., are suitable for a range of building constructions.

Sizes

It is difficult to give a clear definition of timber poles in relation to logs. The diameter of poles for building purposes may be up to 150 mm; however, in some agricultural buildings with large spans, diameters up to 300 mm have been used.

2. HARVESTING, SEASONING AND PRESERVATION

Harvesting

The felling of a tree can be a difficult and dangerous operation and should be carried out carefully. The felling methods most used are:

○ axe

○ power chainsaw.

With experience, it is possible to control the felling direction of the tree and various methods are used to achieve this. In most cases trees will have to be felled according to their lean. Whether axe or chainsaw is used for felling, the decision is made as to which direction the tree is to fall in, and the undercut should face this direction. An undercut is made by removing a segment from one side of the trunk (see Figure 2.1).

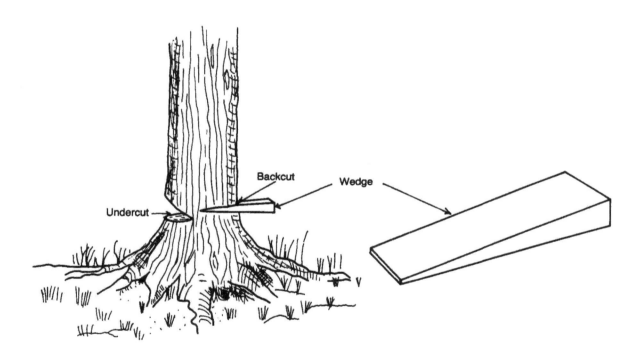

Fig. 2.1 Use of wedge.

After making the undercut, which should go in as near as possible to the centre of the tree, a second cut is made on the opposite side. The second cut is called the back cut and it should be about 100-200 mm above the undercut. Both cuts should open at an angle of between 35°-45°, again very much like a segment. At some point, while the back cut is being made, the amount of wood left in the middle — the hinge — will be too little to support the tree, and so it will fall.

An important aid in felling is a wedge (or sometimes a felling lever), to push the tree into the felling direction when the back cut has been terminated, as necessary. Once the tree starts falling, the operator must retreat backwards along one of the escape routes and watch out for falling branches and other debris.

When a powered chainsaw is used, it is very important to be aware of the safety aspects. Therefore, unless the operator is fully trained, it is not advisable to use a power chainsaw.

Once a tree has been felled, the crown or top of the tree and all the branches have to be cut off.

Felling of trees using chainsaws is explained in detail in the FAO publication *Chainsaws in tropical countries* (FAO Training Series 2).

Debarking

Removal of bark is a very essential step in the preparation of round timbers for treatment, as bark is virtually impermeable to liquids. This does not apply to sap displacement processes such as the Boucherie method, which will be discussed later.

Before use, poles are sometimes debarked, or bark is sawn off and discarded, along with some outer sapwood. There are several ways of removing the bark. Though mechanized peeling gives a smooth, even roundwood, it has the disadvantage of removing too much wood from uneven logs and in addition, it uses expensive machinery. Therefore, in the case of poles, care must be taken to avoid removing too much wood, as this reduces the thickness of the treatable sapwood.

The easiest method of debarking is adzing or by the use of a debarking spud. A debarking spud can be made from a piece of iron approximately 15 mm in diameter and 1350 mm in length (Figure 2.2). The handle is made of round mild steel, to which is welded a short piece of truck spring. The truck spring section is joined to the round mild steel handle by welding. Pieces of rubber tubing are slipped over the round bar to make rubber handles for holding the spud. The blade of the truck spring normally has a curved edge line and this is used to cut the bark and to slip in under the loosened bark to pry it off the pole.

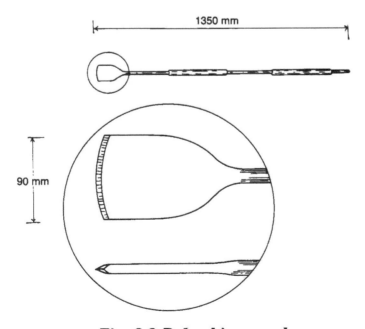

Fig. 2.2 Debarking spud.

3

Peeling and debarking spuds can also be made from a garden spade on which the cutting edge is no longer good, which can be cut down to 100 × 200 mm in size to make a very handy pole peeler (Figure 2.3). The edge of the blade should be ground to a suitable angle, so that it will not tend to cut into the pole.

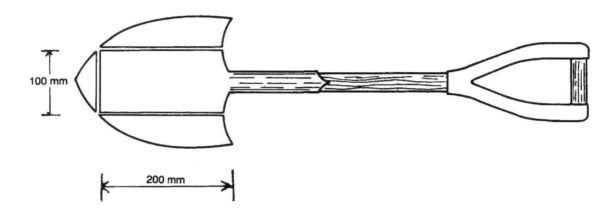

Fig. 2.3 Debarking spud made from an old spade.

When working with smaller material, draw knives may be used very effectively. These are most used during that period of the year when the wood is not in condition to peel, but must be removed by shaving or cutting off.

The blade of a draw knife can be made from 8 × 40 mm tool steel. This could be from a discarded circular saw blade or lightweight car spring (Figure 2.4). The handles are slab pieces of wood held in place with truck brake lining rivets.

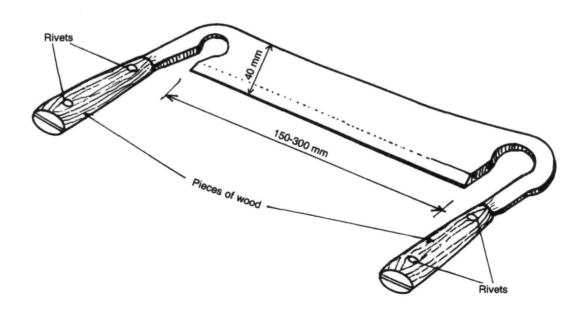

Fig. 2.4 Draw knife.

Seasoning or drying

After debarking, drying may be essential, to reduce shrinkage in use and to prevent deterioration through splits, checks, distortion and decay. It is also a requirement that timber should be seasoned to a moisture content below 25 per cent, before applying a preservative under pressure, or in hot and cold bath. Small poles of the fence post size can reach this stage in a matter of months or even weeks, in dry, warm weather. The period of drying depends upon the species, size and surroundings.

Good stacking is very important to obtain reasonably fast and uniform drying. When stacking, the following points should be remembered:

○ The seasoning site should be in an exposed elevated position with good drainage. It should be open to the prevailing winds.
○ The seasoning area should be kept free of weeds and other vegetation, as well as bark, sawdust and rejected wood, which could be a source of fungal infection.
○ The bottom layer of poles should be raised off the ground at least by 300 mm.
○ Poles should be open crib stacked as shown in Figure 2.5 with adequate space between each piece. Individual stacks should be well spaced, with at least one metre between stacks.
○ The use of rain-shedding cover is highly recommended, as timber which is continually exposed to rain during the drying period is particularly at risk from fungal infection.

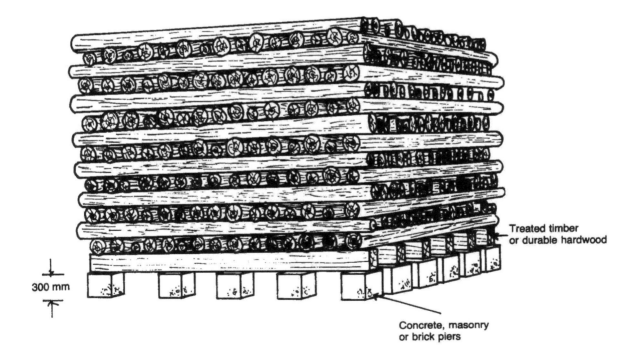

Fig. 2.5 Drying stack.

5

Preservation

Timber preservation in a practical sense refers to the improvement of the natural durability of the timber by treatment with chemicals that are toxic to insects, fungi and other agents of decay; unfortunately, most of these are toxic to human beings.

There are a few species which, in their heartwood, are durable without treatment. Unfortunately, all timber poles contain an outer band of sapwood, which is usually perishable and often amounts to a considerable proportion of the total volume (see Figure 2.6).

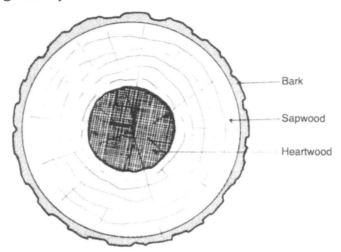

- Bark
- Sapwood
- Heartwood

Fig. 2.6 Cross-section of a pole.

It follows, therefore, that round timber should preferably be treated with a preservative against decay, to promote a longer service life; but proper design is just as important in ensuring the life of a timber structure as is timber preservation. Good detailing in design, such as elimination of all pockets of water, provision of ventilation, etc., minimizes many of these hazards. As far as possible, timber poles should be located so that their ends are not directly exposed to rain, since end grain is a easy access for water. Simple rain caps on exposed beams, pile tops or transmission poles can often extend their service life (see Figure 2.7).

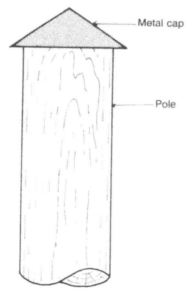

- Metal cap
- Pole

Fig. 2.7 Use of metal cap.

6

Preservatives
Preservatives may be divided into three main classes and they are:
O oil preservatives
O water-borne
O solvent or organic

Oil preservatives
The most common oil preservative is creosote or creosote mixtures. Coal-tar creosote is an oil obtained from the distillation of coal tar. Its colour varies from black to brown. Where the cost of creosote is high, or where absorption is heavy and cannot be satisfactorily controlled, it is economical to dilute it with a cheap petroleum oil, which allows a better distribution of cresote without unduly increasing the initial cost.

Water-borne Preservatives
There are two types of water-borne preservatives, non-fixed water-borne and fixed water-borne type.

Among the non-fixed water-borne preservatives are borax and boric acid, sodium pentachlorophenate, copper sulphate and zinc sulphate. However, most commonly used are the boron compounds, which are applied by the dip diffusion method.

Although they are effective against fungi and insects, boron preservatives are not fixed and hence timber treated with boron salts cannot be used in contact with the ground or in other wet conditions. The preservatives are, however, well suited to the treatment of timber for use above ground inside buildings.

The fixed water-borne preservative type is intended mainly for external use and contains salts which serve to fix the preservative chemicals in the timber and render them non-leachable. The most prolonged and effective protection has been achieved by the use of carefully balanced mixtures of copper, chrome and arsenic salts called CCA Salts. In addition to CCA Salts, there are other similar mixtures copper/chrome/boron, fluorine/chrome/arsenic, etc.

The biggest criticisms against these salts are the poisonous nature of most of them, especially compounds containing arsenic. **Hazards of these and other preservative chemicals should be studied before attempting to use them.**

Organic solvent preservatives
These consist of toxic organic substances dissolved in oil or spirit, which evaporate and leave the dissolved chemical in the wood. These organic solvents have low viscosities and are able to penetrate rapidly into dry wood, so that they are particularly suitable for use in preservative formulations that are designed for superficial application by brush, spray and immersion. Common fungicides for organic solvent preservatives include pentachlorophenol, lindane, tributyl tin oxide (TBTO), copper napthanate. It should be noted that some of the above preservatives are banned in many countries, because of health hazards.

Preparation of material for treatment
If satisfactory results are to be obtained in the treatment of wood, it is essential that the material be suitably prepared for the subsequent impregnation of preservatives.

Only very few processes give best results when applied to green timber, such as the diffusion and Boucherie processes, the latter even requiring the bark to be retained. However, other treatment methods can only be effective if the wood is correctly prepared for treatment. Also, it is extremely important that any cutting, boring or machining operations should be carried out before preservative treatment is given.

Methods of impregnation

Hazards from wood preservatives can be avoided if proper precautions are taken. Toxic chemicals from these preservatives can enter the human body by inhaling, swallowing or by contact with the skin. Information on health and safety procedures are given in the *Wood Preservation Manual* (FAO Forestry Series No 76, published by the Food and Agriculture Organization of the United Nations).

Brushing and spraying

The simplest method of applying a preservative is brushing. Except for the sapwood of more absorbent timbers, this results in little more than skin deep penetration and the protection afforded by one application is slight. Protective clothing, masks and gloves should be worn when carrying out preservative treatment.

Spraying offers a more liberal and effective covering of the timber than brushing. The possibility of the preservative penetrating into holes, cracks, splits, etc., is more in spraying. Tar oils, if they are too viscous, should be heated before applying. Brushing can be repeated at regular intervals, depending on the environment to which it is exposed. Care should be taken to brush any exposed grain with sufficient preservative. It is a must to use gloves and wea protective clothing and to wear a mask over the nose. Preservatives should not be inhaled and should not come into contact with any part of the body.

Dipping or steeping

By allowing timber to soak in a preservative, penetration is increased, but the rate is slow and is, of course, influenced by the permeability of the timber, the resistant timbers being virtually unaffected even after a few months' soaking.

Hot and cold or open tank method

Next to pressure treatment, which will be discussed later, this offers a very satisfactory method of impregnation. In this process, seasoned timber is immersed in a bath of preservative (normally creosote), which is heated to 82-93°C, or as near to this temperature as conditions will allow and held at that temperature for a few hours, after which it is allowed to cool while the timber is still submerged in the liquid. Sometimes, the cooling is done by quickly transferring the timber from the bath to a cool bath of the preservative. Absorption takes place during cooling.

When only butts of fencing and gateposts are treated, any steel tank or drum that is deep enough can be used. Heating can be done using a kerosene pressure burner protected by a sheet-iron wind shield (Figure 2.8a). Alternatively, the drum can be heated using a covered fireplace, as in Figure 2.8b; it is simple to construct and removes any likelihood of the preservative boiling over onto the fire.

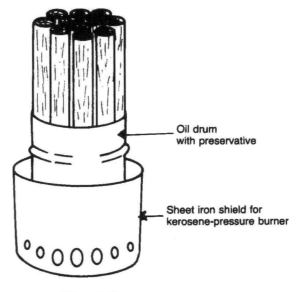

Oil drum
with preservative

Sheet iron shield for
kerosene-pressure burner

Fig. 2.8a.

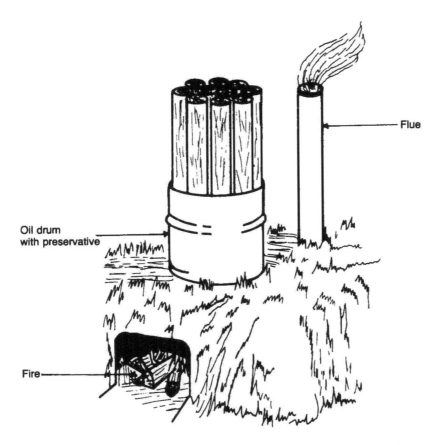

Flue

Oil drum
with preservative

Fire

Fig. 2.8b Hot and cold method using oil drums.

The depth should be sufficient to allow at least 200 mm of treated post to remain above ground level when erected. While the preservative is still hot and before the posts are removed, the tops should be swabbed over with a long-handled brush. However, above ground timber will decay and obviously full-length treatment is the best.

This process is simple and the plant required is cheap, compared with a pressure plant. Figure 2.9 shows a typical arrangement to treat poles, using 200-litre oil drums by cutting them in half and welding them lengthwise and then reinforcing them with angle iron. It is advisable to have a brick foundation, leaving space for an external firing system, whilst minimizing the risk of the preservative catching fire.

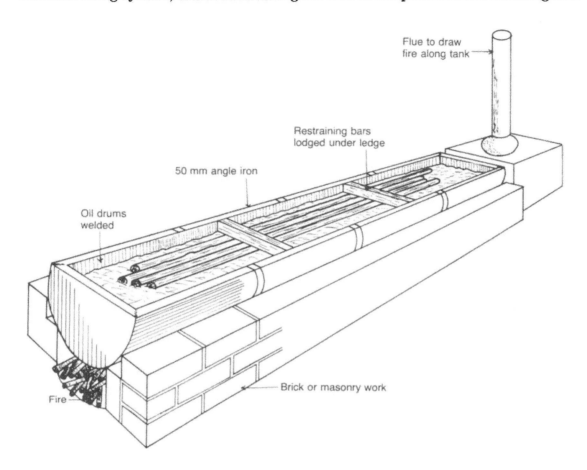

Fig. 2.9 Treatment tank made out of welded oil drums.

The recommended preservative is creosote or creosote fuel oil mixture 50:50. Sometimes, used crank case oil could be used with creosote. However, care should be taken not to allow the the mixture to become a sludge. Once it becomes a sludge, it may not be possible to dissolve the sludge: by experimenting with a sample solution before making the treating solution, it is possible to work out the maximum amount of crank oil that can be mixed with creosote. There is a tendency for the sapwood of some timbers to over-absorb the preservatives. Absorption may be controlled by heating up the preservative a second time, after it has been allowed to cool and removing the timber after one to three hours before cooling starts again.

Sap displacement method

The sap displacement method can only be applied to round timbers in green condition and uses the hydrostatic pressure due to gravity to force the preservative from the butt end of the round timber. The preservative used is normally a water soluble salt such as CCA. In this method, a cap is fitted to the butt end of a freshly cut pole and then one end of a flexible tube is connected to the cap and the other end to a tank containing the preservative at a place as high as possible (see Figure 2.10). Using a manually operated air pump connected to the top of the oil drum can give a significant pressure to force the preservatives into the timber.

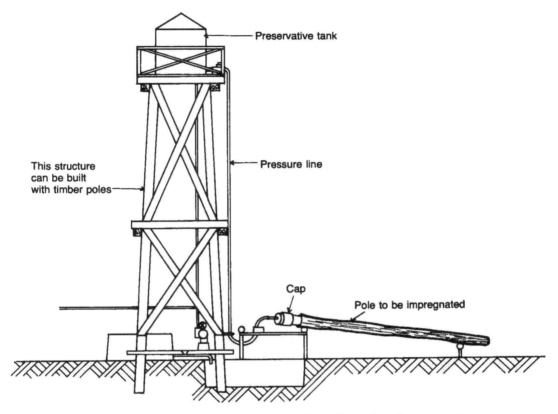

Fig. 2.10 Typical arrangement for Boucherie process.

The caps mentioned above are connected to the pole using old tubes of vehicle tyres (see Figure 2.11).

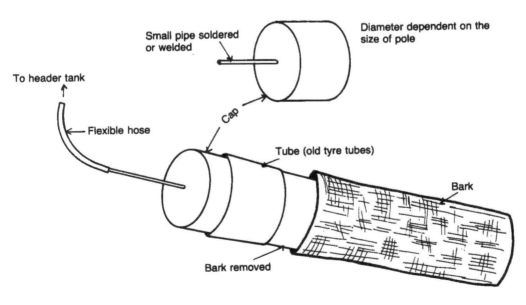

Fig. 2.11 Cap for pole end.

A variation of this method using polythene bags is shown in Figure 2.12.

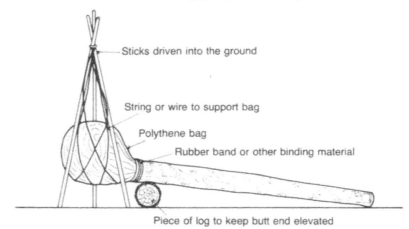

Fig. 2.12 Sap displacement method.

Sap replacement method

This process relies on the upward movement of water or 'sap' which takes place in the sapwood of the growing tree and it is most applicable to young saplings. The young sapling is cut, the bark removed and the bottom end of the pole put in a vessel containing the preservative solution and it is kept upright. This has to be carried out preferably less than 24 hours after felling. The solution, usually CCA, is drawn up to replace the sap as it evaporates from the pole The branches and leaves can be left intact to increase the rate of flow, but they will quickly die. Leaving a greater length of pole than is finally required should increase the flow rate. The pole is left in the bucket or trough until no more solution is absorbed. The duration depends on the size of the pole and atmospheric conditions, and normally it takes two to three days.

A plastic cover tied around the pole and the bucket is needed to prevent evaporation of the solution, or dilution by rainwater (see Figure 2.13).

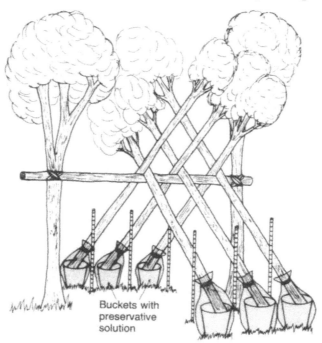

Fig. 2.13 Sap replacement method.

Both methods, sap replacement and sap displacement, give a greater concentration of the preservative at the butt end and this is welcome, especially in the case of poles going into the ground, which are most vulnerable to decay.

Diffusion process

Poles treated by the following method can only be used in a dry condition and not in situations where the timber is going to be wet, or in ground contact. Freshly felled poles are debarked and painted with the preservative in paste form on the surface of the timber. Then they are covered with a polythene sheet or a tarpaulin, to prevent any moisture loss (see Figure 2.14).

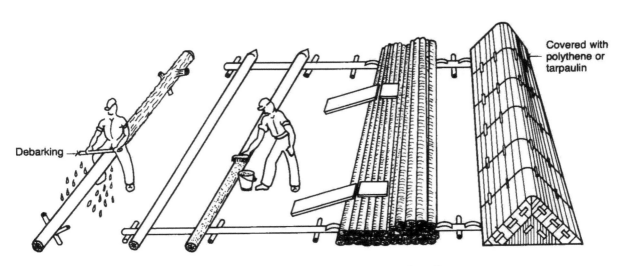

Fig. 2.14 Diffusion treatment of poles.

This method may be useful in treating half round timbers which are not going to get wet during service. The preservative used is a paste containing 20 per cent concentration of borax and boric acid (50:50).

NOTE: The treatment methods quoted above are explained in great detail in *FAO Wood Preservation Manual*, published by the Food and Agriculture Organization of the United Nations, Rome.

Pressure processes

The most successful method of preservative treatment of wood is the use of pressure impregnation of the wood. This is done in specially constructed plants where the timber poles are treated under pressure in a closed steel cylinder. This method involves expensive equipment and so is not discussed in this document.

3. PROPERTIES

Round poles used for transmission and distribution lines and buildings are specified by species, moisture content, grade and dimensions, and their properties vary from species to species.

Normally, the poles are assigned one stress grade higher than the highest grade of sawn timber, because the occurrence of natural defects is compensated for by the inherent strength of round poles (see Chapter 1, Advantages of pole construction). A convenient method of assigning stress grades to round timbers has been adopted in the Australian Standard AS1720.

Another consideration for poles is the effect on strength by the removal of bark. If the bark is removed by adzing, i.e., manually by an adze, spade or machete, then the damage done to the swellings in the trunk around knots is less. If the bark is removed by shaving, the pole will have a smooth profile; but some strength will be lost because of the removal of the natural reinforcement around the knots, exposing the sloping grain. The degree of damage depends on the amount of swelling initially present. A multiplying factor to compensate for shaving or trimming is given in the New Zealand Standard NZ1603: 1981 (see Table 1).

Table 1 Modification factors

Bending stress	0.85
Compressive stress	0.92
Shear stress	1.00
Tensile stress	0.85
Modulus of elasticity	0.95

There is nothing unconventional in the design of pole-frame buildings. Whether round timbers are used as simple structural members, that is as poles or piles, or as elements of a composite structure, the design procedure is the same as with sawn timber. Round timbers are assumed to be in green or dry condition, according to their moisture content at the time of fabrication or installation and in service, except that timbers in ground contact are in all cases assumed for design purposes to be in the green condition at the ground line.

4. POLE-FRAME CONSTRUCTION

Poles, when used for the uprights and set in the ground, serve as both foundations and structural members. Their use leads to a very easily erected and rigid framework to support roof and walls and, if necessary, part of the floor as well.

In pole frame construction, a bracing system may be required at the top of a pole, in order to reduce bending moments at the base of the pole and to distribute loads. The design of buildings supported by poles without bracing requires a good knowledge of soil conditions in order to eliminate excessive or sideways deflection. In some cases, the inherent stiffness of poles properly embedded enables large, open-type buildings to be erected without diagonal bracings between the uprights.

Two distinct types of pole-frame system can be used for house construction. One is known as a 'pole platform' and the other as a 'pole frame'. The 'pole platform' is the simplest form and consists of embedded poles supporting a platform on which a conventional light timber-frame house can be constructed (Figures 4.1, 4.2 and 4.3).

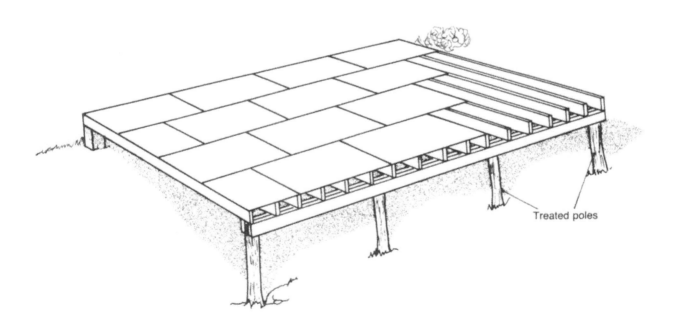

Fig. 4.1 Platform construction using poles with sawn timber.

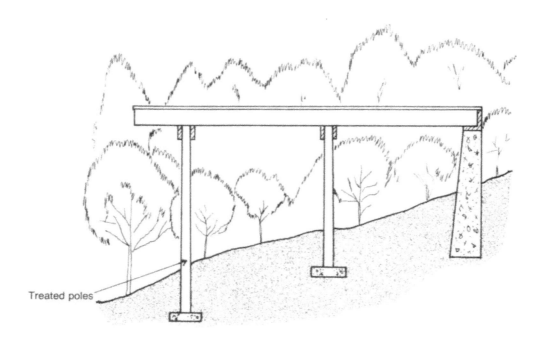

Fig. 4.2 Platform construction.

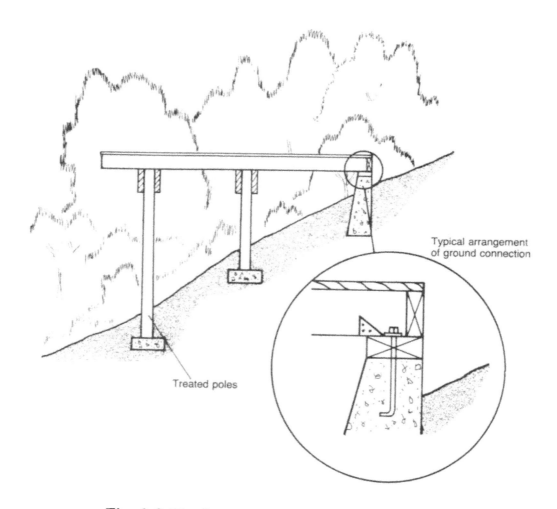

Typical arrangement
of ground connection

Fig. 4.3 Platform construction on a slope.

16

Figure 4.2 shows the simplest way to convert a hillside lot into a buildable site is with a platform if the poles can be embedded to the required depths. If the soil prevents the poles from being embedded deeply enough to resist horizontal forces, the platform should be tied to the hillside as shown in Figure 4.3. This may be applicable to rocky hillsides.

The 'pole frame' is more commonly associated with pole houses and consists of a pole extending from the ground through to the roof, to which cross beams are connected to support the floor (Figure 4.4).

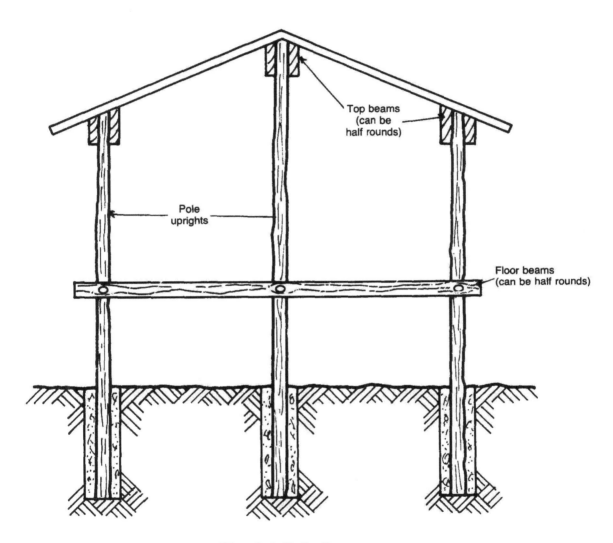

Fig. 4.4 Pole frame.

The pole frame can be braced either with rod bracing as in Figure 4.5, or with wood bracing as shown in Figure 4.6. A simple method of carrying out rod bracing is to fit the rods through drilled holes in the poles, flood with preservative and then to fix with nuts and washers. The size of the rods used is usually 18-20 mm diameter.

In wood bracing, the braces are either nailed or bolted, and form a connection from the pole to a position between the floor beams, as shown in Figure 4.6.

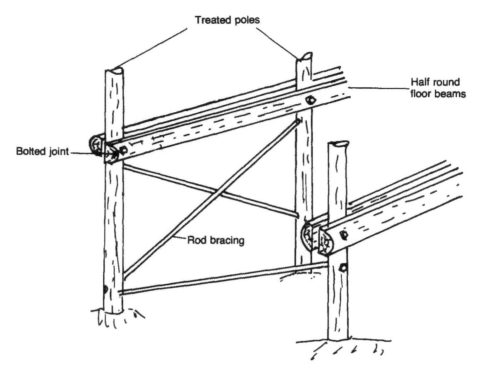

Fig. 4.5 Bracing using iron rods.

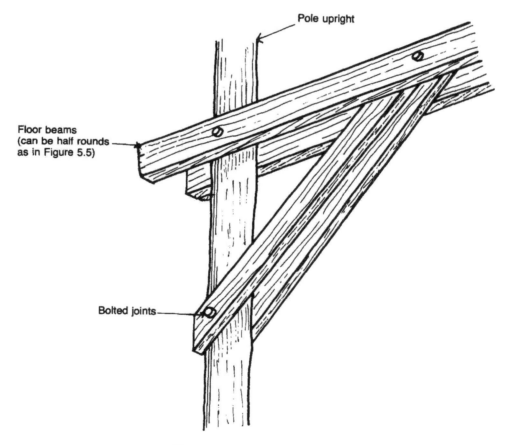

Fig. 4.6 Bracing using timber.

Walls can also be used to act as bracing, as discussed in Chapter 6.

18

5. FOUNDATIONS AND FLOORS

Foundations

Embedding the poles properly is crucial to the strength of a building's frame. When uprights in a pole building depend on ground support, they must be set to a sufficient depth to hold them securely against lateral wind forces on the roof and walls, and also to prevent them from being uprooted by uplift wind forces developed on the structure. Consideration must be given to likely variations in the bearing strength of soils from one part of the site to another and from the wet to the dry season.

If the poles resisting the lateral load cannot be adequately embedded in rocky, sloping terrain, the floor should be tied into the hill as shown in Figure 4.3, or alternative bracing methods should be used.

Low bearing strength soils may be effectively stabilized by the addition of cement. This should be in the proportion ranging from 1 part cement to 5 parts soil to be used at the bottom plate, to 1 part cement to 10 parts soil for the remainder of the backfill. Thorough mixing of soil and cement is necessary before placing and consolidating it around the pole. Lime in similar proportions can be used instead of cement, provided the soil is not very sandy. Water should be added, if necessary, but only to the extent that it slightly moistens the mixture. In addition, proper consolidation is important and it should be well rammed in layers of not more than 150 mm thickness of the loose soil. It is also possible to backfill with concrete or with a graded mixture of gravel and sand, or crushed rock.

For pole structures where vertical loads are heavy, a concrete punching pad at the bottom of the pole hole may be necessary to spread the load (Figure 5.1).

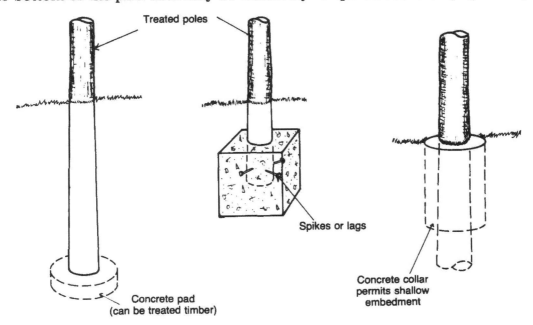

Fig. 5.1 Typical foundations for poles.

19

This can also be achieved using treated round and half-round poles, as shown in Figure 5.2.

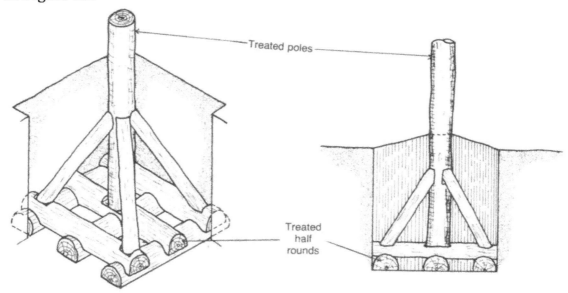

Fig. 5.2 Foundation using treated half rounds.

It is recommended that at least the part of the pole embedded in soil should be treated with a preservative if the pole is not a naturally durable timber. It is also possible to isolate the timber poles from the ground, using concrete footings and steel brackets (Figure 5.3).

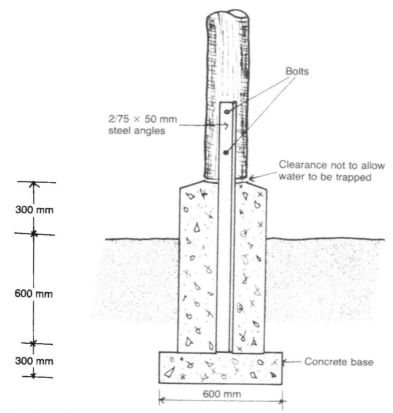

Fig. 5.3 Typical example of the use of steel bracket.

20

Floors

As mentioned earlier, floors can be constructed as shown in Figure 4.1, 4.2, 4.3 or 4.4. It is generally better to frame the pole structure with a pair of beams or girders (Figure 5.4), one on each side. The floor beams could rest on the above girders and the floor could be constructed on the floor beams. The material for the floor could be sawn timber, plywood, or even half rounds. Figure 5.5 shows where half rounds are used for girders as well as the floor itself.

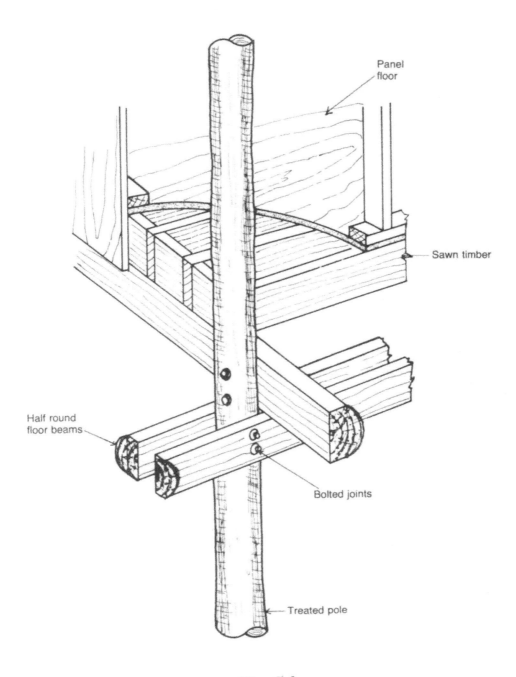

Fig. 5.4.

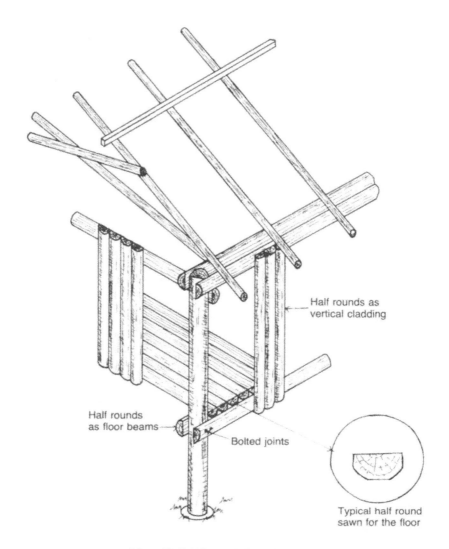

Half rounds as
vertical cladding

Half rounds
as floor beams

Bolted joints

Typical half round
sawn for the floor

Fig. 5.5 Use of half rounds.

An alternative floor framing system is shown in Figure 5.6.

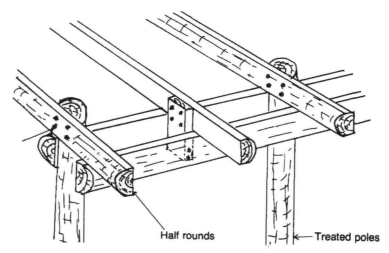

Half rounds

Treated poles

Fig. 5.6 Alternative floor framing.

The actual floor can be half rounds, sawn timber planks, or structural plywood.

6. WALLS

Walls themselves can be used to brace the poles if they are properly designed and installed. The braced wall, rigid under the house loads and properly connected to the poles, can prevent the poles from moving. Walls can be constructed out of sawn timber, plywood, or even half-rounds (Figure 5.5). Sawn timber can be used either in tongue and grooved form, or overlapped (Figure 6.1).

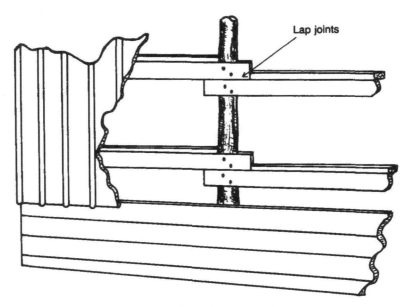

Fig. 6.1 Timber walls.

If plywood is used as sheathing, it should be fixed on stringers (Figure 6.2).

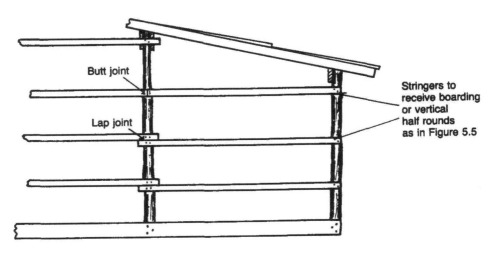

Fig. 6.2 Framing for walls.

As mentioned earlier, walls can be used to brace the poles. The nails, plywood, sizing and tie-downs are all factors in the proper use of this method of bracing. This kind of construction involves additional carpentry where the pole meets the walls. A shear wall utility room is a possible way to brace the structure, as shown in Figure 6.3.

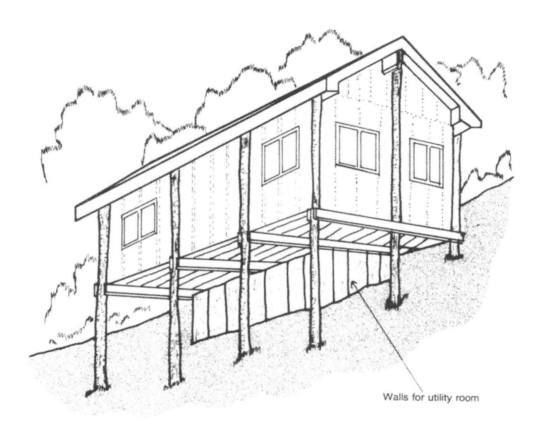

Walls for utility room

Fig. 6.3 Bracing using walls.

7. ROOFS

Most of the following terms are used in connection with roof construction and explained in brief for easy understanding.

Roof covering — the materials laid or fixed on a roof to protect the building. The materials used are: slates, fibre-concrete tiles, burnt clay tiles, asphalt felt, corrugated sheets, wooden shingles and thatch.

Common rafters — similar to beams, but inclined: the distance apart depends upon the covering material.

Span — usually taken to be the clear distance between the internal faces of the walls supporting the roof. The effective span is the horizontal distance between the centre of the supports.

Rise — the vertical height measured from the lowest to the highest points.

Pitch — the slope or inclination to the horizontal, expressed either as rise/span, or in degrees (see Figure 7.1). It varies with the covering material in accordance with Table 7.1, which gives the minimum pitch (see *Building Construction* by WB McKay).

Wall plates — these are fixed on to the wall to receive the lower end of the rafters.

Purlins — horizontal members providing intermediate supports to common rafters.

Table 7.1 Minimum pitch requirements

Covering material	Rise (mm in 100mm run)	Minimum Pitch	Angle
Asphalt felt, corrugated sheets	10	1/160	6°
Slates	50	3/10	30°
Plain tiles and thatch	100	1/2	45°
Interlocking tiles	50	3/10	30°
Shingles	100	1/2	45°

Roof structure of pole buildings can be made out of pole timbers, pole trusses, or out of sawn timber. The sizes of timber members depends, in addition to the span, on the weight of the covering, live loads such as wind, etc.

Table 7.2 gives the average weight of some roof covering materials.

25

Table 7.2 Approximate weight of roof-covering materials

Material	Weight kg/m²
Asphalt felt	3-4
Corrugated iron	12
Corrugated cement	17
Wooden shingles (depending on the type of timber)	6-10
Thatch	34
Slates	43
Clay tiles	35-40
Stone slabs	80-90

The roof-cover can be tiles, sheets, etc. The poles used as principal rafters can either rest on top of the pole column, or be bolted onto the side. Figure 7.2 and Figure 7.3 show two methods of fixing rafters to pole columns.

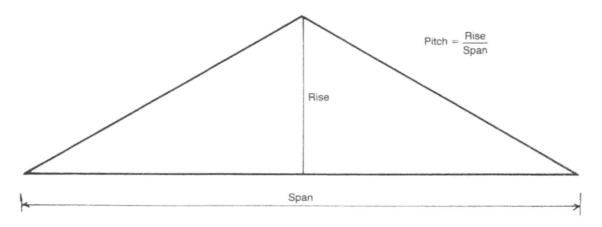

Pitch = $\dfrac{\text{Rise}}{\text{Span}}$

Fig. 7.1 Triangular frame.

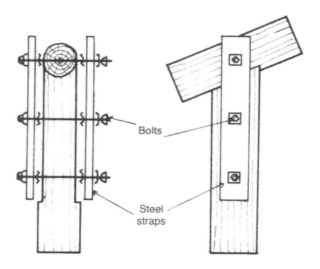

Fig. 7.2 Rafter secured with poles.

26

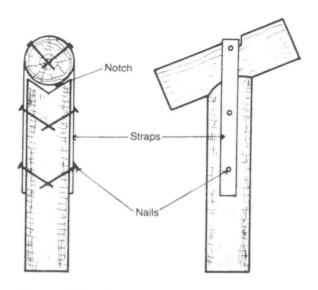

Fig. 7.3 Rafter nailed and strapped.

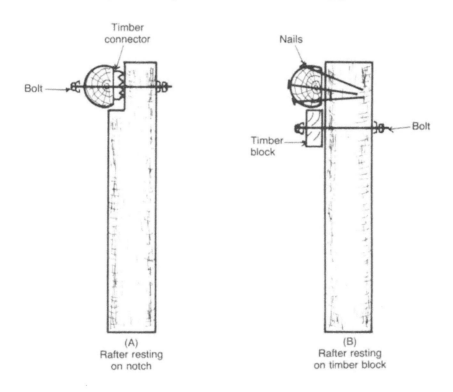

Fig. 7.4.

Both methods require the height of poles to be measured very accurately; but to avoid this, vertical poles could be cut, so that they are all level, after they have been securely erected. If metal straps are used, it is advisable to use corrosion-resistant material. This applies to nails too. A shallow notch which can be slightly varied to allow for any unevenness in the size of the poles, providing a reasonably straight roof line, should be made on the underside of the rafter to get an even bearing. The top of the rafters, when necessary, should be rounded off at the point where the metal strap is secured. When bolts are used, the diameter of the timbers should not be less than 100 mm. In this case, the sides of poles should be squared with a saw before bolting on the side of the straps.

27

Figure 7.4 shows another method of fixing rafters to poles. The poles can be notched and then supported by a timber block. The block is nailed before bolting to the pole, so that the level of the roof can be adjusted before the hole is drilled or bolts inserted. If a timber connector is not used, it is important to make sure that the rafter gets maximum support by resting squarely on the block or shoulders of the notch. The joints made with timber connectors are discussed in Chapter 8.

A method for fixing the tops of the rafters is given in Figure 7.5. As shown in this Figure, tops of the rafters can be sawn off at an angle, so that they could face each other where they rest and be nailed into place.

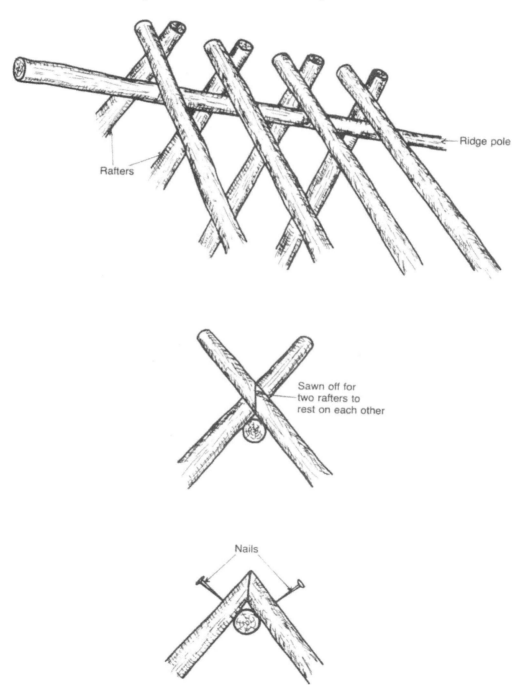

Fig. 7.5 Rafter and ridge pole connection.

Roof structure can be constructed out of roof trusses which may be of pole timbers or sawn timber. Figure 7.6 shows a scissor truss which can be made out of round timber. It is very important to make sure that the trusses of this type should be cross-braced properly, to avoid them collapsing sideways.

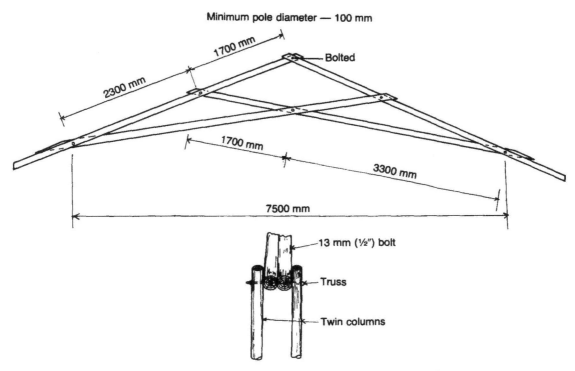

Fig. 7.6 Scissor truss and connection to column.

Figure 7.7 shows a halved pole truss developed by the Building Research Establishment in UK. The joints are made out of nails. However, they can be made using wooden pegs.

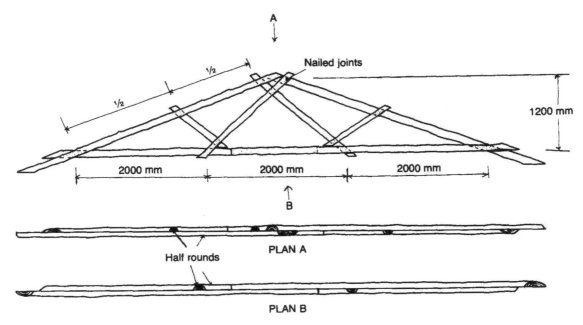

Fig. 7.7 Truss using half rounds.

29

When poles are used in roof structures with masonry or adobe buildings, it may be necessary to have a wall plate embedded on top of the wall, as shown in Figure 7.8, for the rafters to rest on.

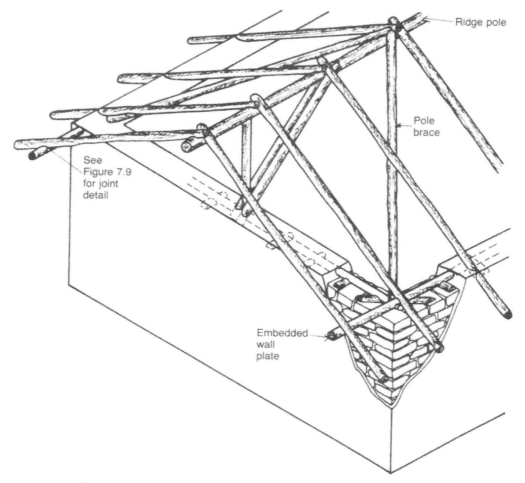

Fig. 7.8 Pole roof structure on a brick wall.

Figure 7.9 shows a method of fixing the rafter to the wall plate by having notches in the rafter and the wall plate (pole) to fit into each other and nailing them into place.

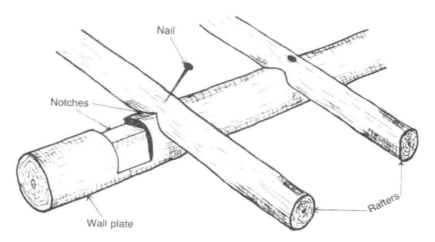

Fig. 7.9 Rafter wall plate connection.

8. JOINTING TECHNIQUES

The most important aspect of a pole timber structure is the joints. The traditional craft methods of connecting the poles (using coir rope, sisal rope, etc.,) do not permit the full strength of poles to be utilized. However, the above methods have to be employed for smaller size poles (less than 75 mm in diameter), as it is difficult to use most other jointing techniques described below.

Connections in round timber members are more difficult, because of their shape. In addition, in the case of fast grown species, especially in the case of young saplings, the timber tends to shrink more, as a result of drying after being cut. Splitting occurs mostly at the ends. Serious end-splitting is not unusual; in some cases it can be reduced in severity if the cross-cut ends are thoroughly coated with a suitable drying retardant soon after felling. The best retardants appear to be petroleum-type waxes, which can be applied by brushing the pole ends generously.

Substantial re-splitting, after cutting of a badly split end of a seasoned log, may be prevented by applying a tight band of metal strapping or wire just inside the position of the new end before making the cross-cut or to use a metal cap promptly and tightly fitted to the end immediately after cross-cutting. In the case of transmission poles, nailed plates are driven into the ends to control splitting.

Design of joints

The timber joints in pole-type structures are similar in design to those for other timber structures, except for the complexity induced by the roundness of the pole. The most common joints are made with nails, bolts and nuts. The roundness need not be a consideration when using bolted connections as in Figure 8.1.

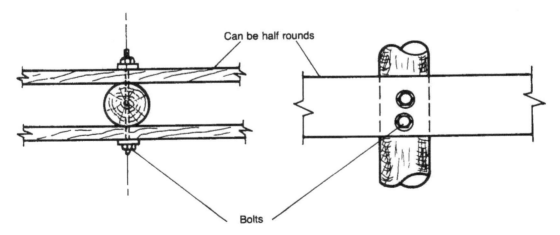

Fig. 8.1 Joint using bolts.

The curve of the pole can be eliminated by having a notch as shown in Figure 8.2 and then the connection made with bolts, nails, gusset plates, or with ordinary connectors. The notch can also form a seal for the beam, to resist vertical loads. The notch can also receive a strap that ties down a beam carried on top of a pole (Figure 8.3).

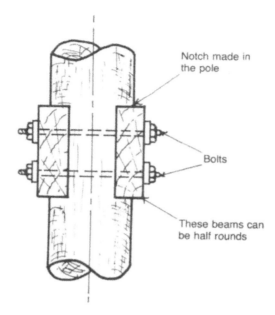

Fig 8.2 Beams resting on notches made in the pole.

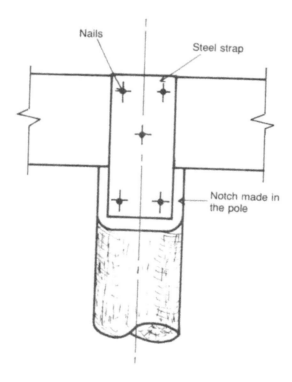

Fig. 8.3 Connection using a steel strap.

Timber connectors

The most suitable timber connector for pole joints is the spike grid. Figure 8.4 shows a single curved spike grid inserted between the pole and the beam, which increases the strength of the bolted connection.

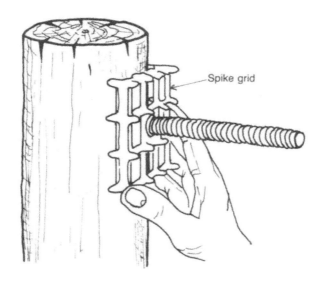

Fig. 8.4 Single curved spike grid.

With the curved side of the grid against the pole and over pre-drilled holes, a high-strength threaded rod is used to squeeze the two timber surfaces together, enough to force the teeth of the spike grid into the grain of both members. The high-strength rod is then replaced with a conventional bolt of the proper size.

It is estimated that each single curved spike grid with 25 mm bolt has a load carrying capacity of 2,000 kg, whilst without a spike grid it is around 1,200 kg.

Jointing techniques using wood or steel dowels

A jointing technique using dowels was developed at the University of Nairobi. These dowels are fitted into pre-drilled holes (Figure 8.5). If the loads are not too great, the dowels could be made out of wood, which does not corrode. However, they should be prevented from slipping out by means of nails or pegs inserted at different angles. Alternatively, holes can be drilled into the ends of the wooden dowels, into which hardwood wedges can be fitted to keep the dowel in place. Thus the hole into which the dowel is inserted can be slightly oversized, to facilitate and speed up work.

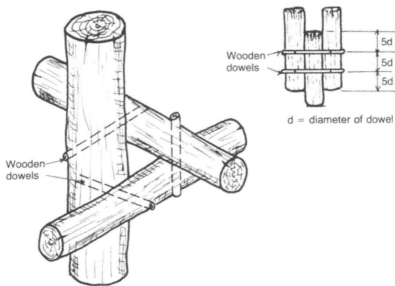

Fig. 8.5 Use of steel/wooden dowels.

33

Where steel dowels are used, to prevent them slipping out of the timber, 10-12 mm deep holes should be drilled into the ends of the dowels, as described above in the case of wooden dowels. With a cross saw cut, the end pieces can then be hammered back like flower petals, holding down a steel washer (Figure 8.6).

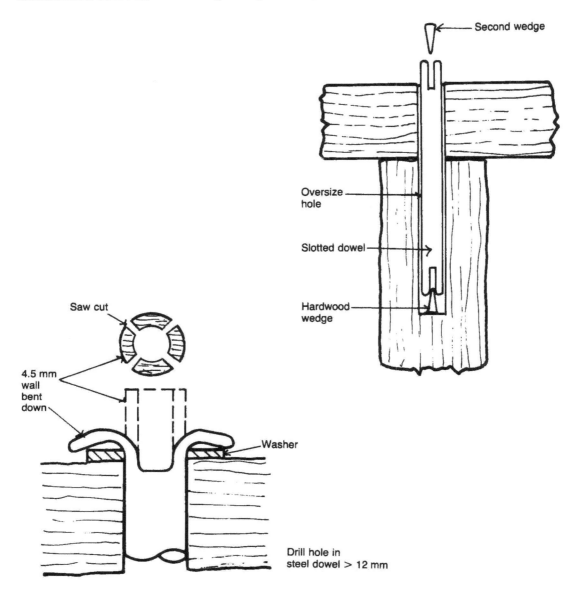

Fig. 8.6 Use of steel dowels.

Joints using metal plate connections

This simple and cheap technique, developed at JPM Parry Associates' Workshops, Cradley Heath, UK, uses thin sheet metal (up to 1 mm thick), cut to the required size and shape, which is wrapped around the joints and firmly nailed on to the timber. The most suitable application of this method is in the prefabrication of pole timber trusses. To ensure uniform dimensions, the trusses are made with the help of templates laid on the ground and held in place by wooden or steel pegs. The poles are placed as accurately as possible on the template, then cut to size and joined together, as described below (Figure 8.7).

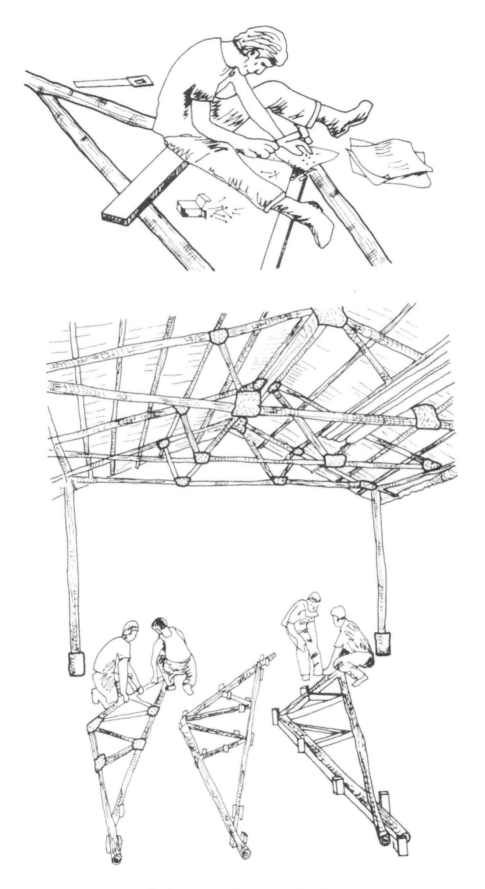

Fig. 8.7 Joints using metal plates.

Joints using flitch plate connections

A method for forming pole timbers into a structural frame using a nailed steel flitch plate, developed at the Building Research Establishment (BRE), Watford, UK, consists of mild steel sheets inserted into longitudinal saw cuts in the timber poles and connected to them by nails driven through the timber and the steel at right angles to the plate (Figure 8.8). Thicker steel plates require drilling, or the use of hard steel nails. Use of 1 mm thick plates can easily be penetrated by normal round wire steel nails without pre-drilling. It has been found it is better to increase the number of 1 mm plates, rather than their thickness. Roof frames that need to be stronger may require flitch plates of larger areas to achieve appropriate design stresses.

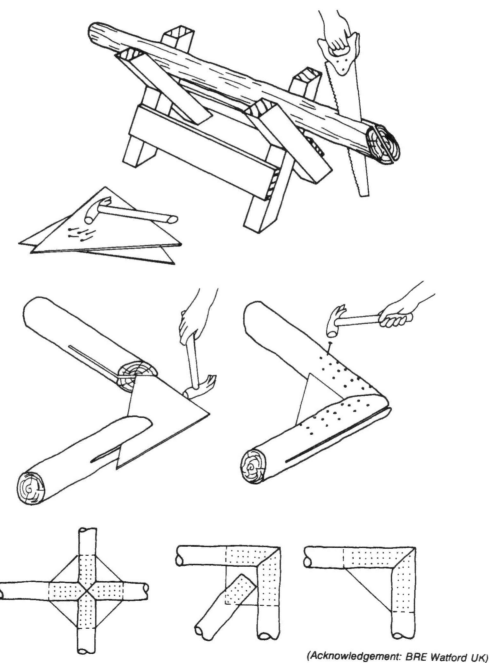

(Acknowledgement: BRE Watford UK)

Fig. 8.8 Joints using flitch plate connections.

Joints using wire lacing

A timber pole joint using a metal plate and wire lacing tool has been developed by the Civil Engineering Department of the Delft University (Figure 8.9). This tool uses galvanized wire up to 4 mm thick and it stretches the wire little by little until the structural parts to be connected are compressed firmly and until the wire lies tightly around the material. The tool can also be used to join two poles, without a metal plate being used (Figure 8.9c). This method is especially useful for small diameter poles, where nailed or bolted joints cannot be used.

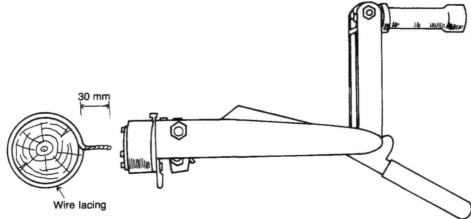

30 mm

Wire lacing

Fig. 8.9a Wire lacing tool.

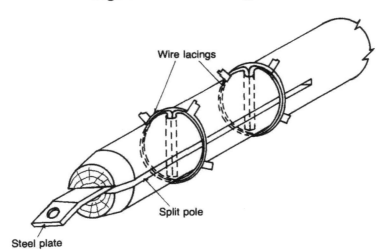

Wire lacings

Split pole

Steel plate

Fig. 8.9b Use of steel plate with wire lacing.

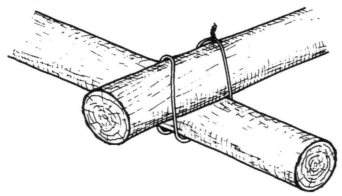

Fig. 8.9c Use of wire lacings.

Other types of joints

Figures 8.10 to Figure 8.13 show other types of round timber connections and these have been reproduced from reference 30 (*Pole Structures* by GB Walford).

Figure 8.10 is a typical bracket made of galvanized steel which is commonly used for the bracing of pole frame and pole platform house construction.

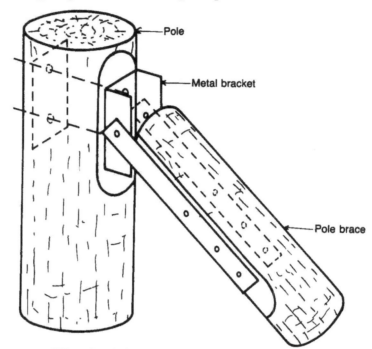

Fig. 8.10 Modern metal bracket.

Figure 8.11 shows pole/rafter joints using U bolts.

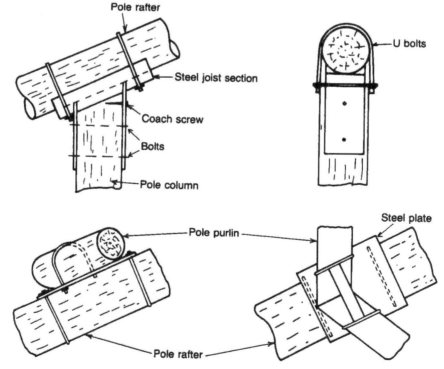

Fig. 8.11 Connections using U bolts.

Typical connections using steel straps are shown in Figure 8.12.

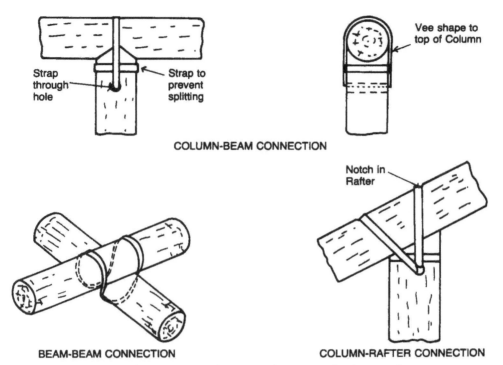

Fig. 8.12 Connections using steel strapping.

Figure 8.13 shows connections using threaded steel rods, where the minimum diameter of the poles used for this type of connection has to be more than 100 mm to avoid the risk of splitting.

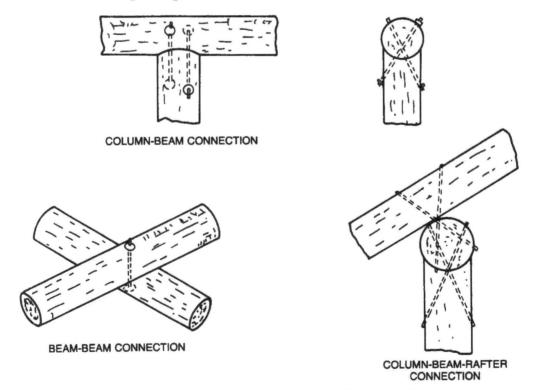

Fig. 8.13 Connections using threaded steel rods.

39

9. OTHER USES

In addition to the uses mentioned earlier, poles are also used for bridge stringers, shelter fences, retaining walls, wharf piles, groynes, agricultural storage buildings. Examples of some of these uses are given below.

Space frame structures

Space frame structures such as domes can conveniently be constructed using wire lacing techniques.

Another method of using short-length pole timber to construct space frames for large covered areas was developed in Sweden by Habitropic. The system is based on special space frame connectors, comprising a cross-component of welded steel, and tail-end connectors with screws, washers and nuts (see Figure 9.1). The poles are all cut to the same length, say 1.5 m, and cut length-wise at both ends with a saw. Holes for bolts are drilled at each end, the steel tail-end connectors inserted in the saw cut and fixed with bolt, washer and nut. After prefabricating all the required poles, they are assembled on the ground, directly below their final position and lifted into place by a pulley system.

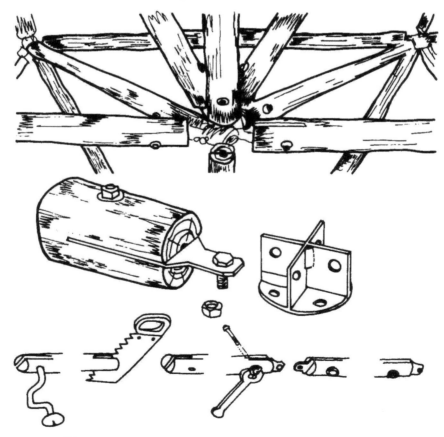

Fig. 9.1 Connections for a space frame.

Retaining walls and groynes

Retaining walls and groynes are employed to hold earth or beach sand. These structures have been constructed using treated timber poles. Figure 9.2 and Figure 9.3 show typical details of a sea-wall and those of a timber groyne respectively.

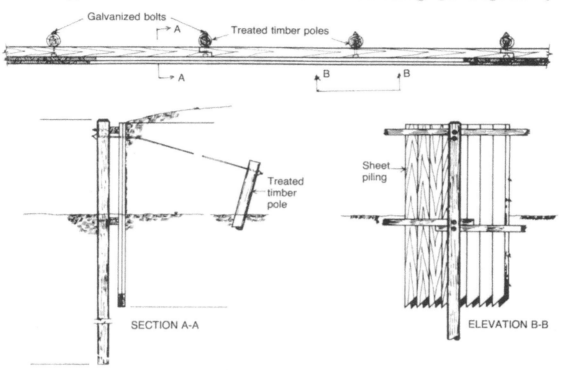

Fig. 9.2 Typical sea-wall or bulkhead.

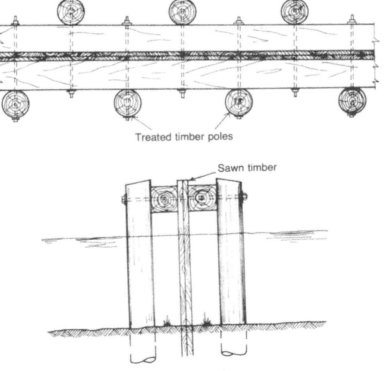

Fig. 9.3 Typical timber groyne.

Bridge construction

Treated timber poles may be used as beams in the construction of timber bridges for rural roads. A sketch of a typical timber pole bridge is given in Figure 9.4.

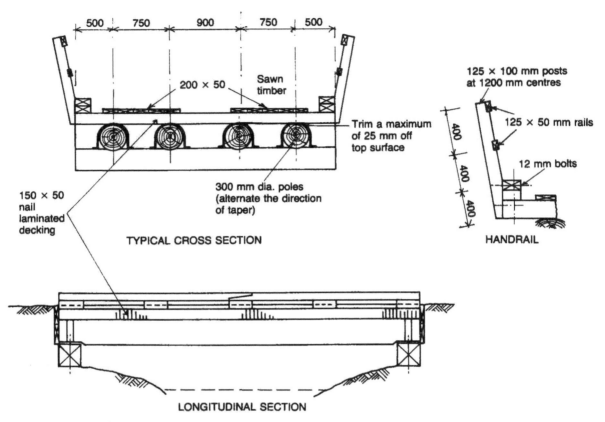

Fig. 9.4 Timber pole bridge (spans 3 to 6 metres).

10. REFERENCES

In writing this document, the author has used not only his own experience, but also much information extracted from the references given below.

1. *How to build storm-resistant structures* American Wood Preservers Institute, 1972. Virginia USA. 24pp.
2. *Pole house construction* American Wood Preservers Institute, 1970. Virginia USA. 30pp.
3. 'New Zealand engineer predicts boom for pole frame construction' Anon. *Australian Forest Industries Journal*, October 1980. 46(9)39-40.
4. *Pole type constructions Pts 1 & 2*. Blakeley J and Nauta F. New Zealand Timber Research and Development Association, Timber and Wood Products Manual. Section 2f-1. Wellington, TRADA, 1973. 6pp.
5. *Efficient engineering with round timbers* Boyd, JD. The Institution of Engineers, Australia, 1961. Vol 33 Nos 1-2. 16pp.
6. *Pole hole construction* Building Research Association of New Zealand. Building Information Bulletin 232. Porirua, BRTANZ.
7. *Pole barn for hardening-off calves at NAC Farm* Buildings Information Centre. Building Report 80. Farm Building Digest, Spring 1976.
8. *Chainsaws in tropical countries* FAO Training Series, Food and Agriculture Organization, Rome, 1980. 96pp.
9. *Pole buildings in Papua New Guinea* Forest Products Research Centre, Department of Forests, Papua New Guinea. 1975. 44pp.
10. *Observations and trials on two methods of connecting pole timbers to form structure frames* Proceedings of an International Conference on Low Cost Housing for Developing Countries, Herbert, M, Building Research Establishment, November 1985. pp 645-654.
11. *Low-cost space frame roof structures*, Habitropic, Birkagatan 27, S-113 39 Stockholm Sweden, 1983.
12. *The Delft wire lacing tool* Huybers, Pieter. *AT News* Vol 5, 1984. Delft Netherlands.
13. *Specification for wood poles for overhead power and telecommunication lines* Indian Standards Institution, India 1962. 14pp.
14. *Poles from forest thinnings for low cost housing* Building Research and Practice. Jayanetti, DL. July/August 1978. CIB Paris. pp250-254.
15. *Wood Preservation Manual* Jayanetti, DL. FAO Forestry Series No. 76, Food and Agriculture Organization, Rome, 1986. 160pp.
16. *Design of timber pole barns* Kelly, M (Reprint). Farm Building Progress, January 1976, (43)13-16.
17. *Building construction (Metric)* McKay, WB. Volume I. Longman Group, London. 162pp.
18. *Low-cost pole building construction* Merilees, D, and Loveday, E, Charlotte, Vermont. Garden Way Publishing, 1975. 102pp.
19. *Round timber in the farm* Ministry of Agriculture, Fisheries and Food, UK. August 1970. 14pp.
20. 'Construction of pole-system housing' *Timber and Wood Products Manual*. New Zealand Timber Research and Development Association (Inc). Section 2f-2. Wellington, TRADA, 1976. 4pp.
21. 'Pole frame buildings' *Timber and Wood Products Manual*, New Zealand Timber Research and Development Association (Inc). Section 2b-1. Timber Impregnation Co (NZ) Ltd, 1976. 32pp.
22. *The New Zealand pole house*. Norton, P ed. Auckland, Hickson's Timber Impregnation Co., (NZ) Ltd, 1976. 32pp.
23. 'Pole-type structures' *A planning guide for various sizes of rafter styles and strength requirements*. Oregon State University Co-operative Extension Service. Pacific Northwest Co-operative Extension Publication 100. Idaho, USA. July 1968. 31pp.
24. *A study of wind stresses in three typical pole building frames* Parker, JF. US Forest Products Laboratory, Research Note FPL-049.

25. *Shanty upgrading* Parry, John and Gordon Andrew (Technical handbook for upgrading squatter and shanty settlements). JPM Parry Associates' Workshops, Cradley Heath, UK. 1987.

26. *Pole building design* Patterson, D. American Wood Preservers Institute, Washington. 6th ed 1969. 48pp.

27. *Pole frame housing and the environment: Notes from the Seminar held at the University of Canterbury, 17 and 18 February 1976*. School of Forestry University of Canterbury, (NZ). 1976. 123pp.

28. *Pole housing design and practice: Notes from the Seminar held at the University of Canterbury, 9 and 10 May 1977*. School of Forestry (NZ), Christchurch University 1977. 199pp.

29. *Engineering design data for radiata pine poles* Walford, GB and Hellawell, CR. New Zealand Timber Research and Development Association, Timber and Wood Products Manual, Section 2b-2. Wellington, TRADA, 1972 5pp.

30. *Pole structures* Walford, GB. Paper presented at the Australia/UNIDO Workshop in Timber Engineering 2-20 May 1983. Melbourne Australia. 12pp.

31. *Code of practice for timber* Design Standards Association of New Zealand (1981). NZ 3603: 1981, New Zealand. 75pp.